Mining in Africa

PAST, PRESENT AND FUTURE.

While every precaution has been taken in the preparation of this book, the publisher assumes no responsibility for errors or omissions, or for damages resulting from the use of the information contained herein.

MINING IN AFRICA

First edition. June 20, 2024.

ISBN: 979-8227718174

Written by Sibusiso Anthon Mkhwanazi.

Table of Contents

Sibusiso Anthon Mkhwanazi

Dedication

Whether you are a scholar, policymaker, industry professional, or concerned citizen, this book provides a comprehensive and thought-provoking exploration of mining in Africa, offering valuable insights into its past, present, and future.

This book is dedicated to the resilient people of Africa, whose rich history, diverse cultures, and unwavering spirit have shaped the continent's mining industry. May their efforts towards sustainable development and inclusive growth inspire generations to come.

Acknowledgments:

I would like to express my sincere gratitude to all the individuals and organizations whose contributions and support made this book possible.

Special thanks to the researchers, scholars, and experts whose valuable insights and expertise enriched the content of this book. Your dedication to advancing knowledge and understanding in the field of mining in Africa is truly commendable.

I am grateful to the African communities, mining companies, and government agencies who generously shared their experiences, perspectives, and data, providing invaluable firsthand insights into the realities of mining operations on the continent.

I would also like to thank the editors, proofreaders, and publishing team for their meticulous attention to detail and dedication to producing a high-quality publication.

Finally, I extend my heartfelt appreciation to my family, friends, and colleagues for their unwavering support, encouragement, and understanding throughout the writing process.

Thank you all for being part of this journey.

Introduction

1. Overview of Mining in Africa

Mining has played a pivotal role in shaping the economic landscape of Africa. The continent is endowed with a vast array of mineral resources that have attracted global interest for centuries. The importance of mining in Africa cannot be overstated, as it is integral to the continent's economic development, providing significant contributions to GDP, export revenues, and employment.

Importance of Mining in Africa

Mining is a critical economic driver for many African countries, acting as a major source of revenue and foreign exchange. The continent is rich in a variety of minerals, including gold, diamonds, platinum, copper, cobalt, and coal, among others. These resources are essential not only for local economies but also for the global supply chain.

1. **Economic Contribution**: Mining contributes significantly to the GDP of several African nations. For instance, in countries like South Africa, Ghana, and Botswana, mining constitutes a substantial portion of national economic output. In South Africa, mining accounts for approximately 8% of the GDP, while in Botswana, diamond mining contributes around 20% of the GDP.

2. **Employment**: The mining sector provides employment to millions of people across the continent, both directly and indirectly.

It offers jobs in various capacities, from mining engineers and geologists to laborers and administrative staff. In addition to direct employment, mining stimulates job creation in related industries, such as transportation, manufacturing, and services.

3. **Foreign Direct Investment (FDI)**: Mining attracts substantial foreign direct investment, with multinational corporations investing in the exploration and development of mining projects. These investments bring much-needed capital, technology, and expertise to the host countries, fostering economic growth and development.

4. **Infrastructure Development**: Mining operations often necessitate the development of infrastructure, such as roads, railways, ports, and power supply. This infrastructure not only supports mining activities but also benefits local communities and other economic sectors.

5. **Export Revenue**: Minerals are among the top exports for many African countries, generating significant foreign exchange earnings. For example, gold is a major export for Ghana and South Africa, while diamonds are critical for Botswana and Namibia. These revenues are vital for balancing national budgets and financing development projects.

Brief History of Mining Activities

The history of mining in Africa is as rich and diverse as the continent itself, dating back thousands of years to ancient civilizations. Over the millennia, mining has evolved from rudimentary methods to sophisticated, large-scale operations.

1. **Ancient Mining Practices**: Mining activities in Africa can be traced back to the ancient Egyptians, who mined for gold and copper as early as 3000 BCE. Gold was particularly valued, and evidence of ancient mining has been found in the Nubian Desert,

indicating the existence of extensive mining operations. The Great Zimbabwe civilization, which flourished between the 11th and 15th centuries, also engaged in mining, extracting gold and trading it with distant lands.

2. **Colonial Era Mining**: The discovery of minerals in Africa during the colonial period (late 19th to mid-20th century) led to a rush of European powers seeking to exploit these resources. Colonial governments and private companies established large-scale mining operations, often at the expense of local communities and environments. South Africa's Witwatersrand Gold Rush in 1886 is one of the most notable examples, leading to the establishment of Johannesburg and transforming the region into a major gold producer.

3. **Post-Colonial Developments**: Following independence in the mid-20th century, African nations sought to reclaim control over their natural resources. Nationalization of mines and the establishment of state-owned mining companies were common, as newly independent governments aimed to ensure that mining benefits were more equitably distributed. However, this period also saw challenges such as political instability, lack of expertise, and inadequate infrastructure, which hindered the growth of the mining sector.

In recent decades, reforms and liberalization of the mining sector have attracted foreign investment and technology, leading to the modernization of mining operations. The introduction of new policies and regulatory frameworks has aimed to balance the interests of investors with the need for sustainable development and local community benefits.

Scope and Purpose of the Book

This book aims to provide a comprehensive examination of the mining sector in Africa, offering insights into its historical context, economic significance, and the multifaceted impacts it has on societies and environments. The scope of the book covers a wide range of topics to give readers a thorough understanding of the various dimensions of mining in Africa.

1. **Historical Perspective**: We will explore the historical development of mining in Africa, tracing its origins from ancient practices to modern industrial activities. This includes an examination of how colonial and post-colonial dynamics have shaped the mining industry.

2. **Types of Mining**: The book will delve into the different types of mining prevalent across the continent, including gold, diamond, base metals, coal, and artisanal mining. Each type will be discussed in terms of techniques, key regions, and economic contributions.

3. **Economic Impact**: We will analyze the economic contributions of mining to African economies, focusing on GDP, employment, foreign direct investment, and infrastructure development. This section will highlight how mining drives economic growth and development.

4. **Social Impact**: The social ramifications of mining will be thoroughly examined, including its effects on community development, labor conditions, and cultural heritage. We will look into both the positive and negative social outcomes of mining activities.

5. **Environmental Impact**: Environmental issues related to mining, such as deforestation, pollution, and sustainable practices, will be discussed. We will also explore the regulatory frameworks and policies aimed at mitigating environmental damage.

6. **Case Studies**: Detailed case studies of major mining countries like South Africa, the Democratic Republic of Congo, Botswana, and Ghana will provide specific examples of how mining affects different regions. These case studies will offer insights into successes, challenges, and lessons learned.

7. **Challenges and Controversies**: The book will address various challenges and controversies in the mining sector, such as the resource curse, corruption, conflict minerals, and economic diversification strategies.

8. **Future Prospects**: Finally, we will look at the future of mining in Africa, considering technological innovations, renewable energy integration, and emerging policies. This section aims to provide a forward-looking perspective on the potential for sustainable and inclusive growth in the mining sector.

The purpose of this book is to inform and engage a wide audience, including policymakers, industry professionals, academics, and general readers interested in the economic and social development of Africa. By providing a detailed and balanced view of mining in Africa, we hope to contribute to a more informed and constructive discourse on the continent's future.

Chapter 1: Historical Context

1. Ancient Mining Practices

Mining activities in Africa date back thousands of years, showcasing the ingenuity and resourcefulness of early civilizations. These ancient practices laid the groundwork for the continent's rich mining heritage, revealing a sophisticated understanding of geology and metallurgy long before the advent of modern technology.

Early Civilizations and Mining

Ancient Egypt:

One of the earliest and most notable examples of ancient mining in Africa comes from Egypt. The Egyptians began mining for gold and copper as early as 3000 BCE. Gold, in particular, held great significance in ancient Egyptian culture, symbolizing eternity and the divine. The gold mines of Nubia, a region south of Egypt, were some of the most famous in the ancient world. The precious metal was used extensively in jewelry, coinage, and as a medium of exchange.

Nubia:

Nubia, known for its rich deposits of gold, was a significant mining region during the time of the ancient Egyptians. Nubian mines provided a substantial portion of the gold that fueled the wealth of the Egyptian empire. Mining in Nubia involved large-scale excavation and the use of primitive yet effective techniques to extract the metal from the earth.

The Great Zimbabwe Civilization:

The Great Zimbabwe civilization, which thrived between the 11th and 15th centuries in what is now modern-day Zimbabwe, also engaged in extensive mining activities. Gold was the primary mineral extracted, and the civilization became known for its sophisticated mining techniques. The gold from Great Zimbabwe was traded along the Swahili Coast, reaching as far as China and the Middle East, highlighting the civilization's role in regional and international trade networks.

West Africa:

In West Africa, particularly in the regions of modern-day Ghana and Mali, ancient civilizations were heavily involved in gold mining. The ancient Ghana Empire, flourishing between the 6th and 13th centuries, controlled vast gold fields and was renowned for its wealth. Similarly, the Mali Empire, under the reign of Mansa Musa in the 14th century, became legendary for its gold production, which significantly contributed to its prosperity and influence.

Traditional Mining Methods

The mining techniques employed by ancient African civilizations were remarkably effective given the technological limitations of the time. These methods evolved over centuries, reflecting a deep understanding of the natural environment and resource management.

Alluvial Mining:

Alluvial mining, or placer mining, was one of the earliest forms of mining practiced in Africa. This method involved the extraction of minerals from riverbeds and stream sediments. Gold particles were separated from the sand and gravel using simple tools such as pans, sluices, and cradles. This technique was particularly common in regions with abundant water sources, such as the rivers of West Africa.

Surface Mining:

Surface mining, including open-pit and quarrying techniques, was widely used to extract minerals located near the earth's surface. Ancient miners dug shallow pits or trenches to access mineral deposits, employing tools made from stone, wood, and later, metal. These surface operations were labor-intensive but allowed for the extraction of valuable minerals like gold, copper, and iron.

Underground Mining:

For deeper mineral deposits, ancient African miners developed underground mining techniques. They excavated tunnels and shafts to reach the ore bodies, using basic tools like chisels, hammers, and wooden supports to prevent cave-ins. Ventilation was a significant challenge, and miners often relied on natural airflow or rudimentary ventilation systems to maintain breathable air in the tunnels.

Fire-Setting:

One of the more ingenious techniques used by ancient miners was fire-setting. This involved heating the rock face with fire and then rapidly cooling it with water to induce cracking, making it easier to break apart. Fire-setting was commonly used in hard rock mining to extract minerals embedded in solid rock formations.

Metalworking and Smelting:

The extraction of metals like gold, copper, and iron required advanced knowledge of metallurgy. Ancient African metallurgists developed sophisticated smelting techniques to process ores. Furnaces were constructed using clay and stone, capable of reaching high temperatures necessary for smelting. The addition of fluxes, such as charcoal, helped to reduce the ores and separate the metals. The resulting metal was then cast into ingots, tools, and ornaments.

Trade Networks:

Mining activities in ancient Africa were closely linked to extensive trade networks. The gold from regions like Nubia, Great Zimbabwe, and West Africa was highly sought after and traded across vast distances. These trade routes facilitated the exchange of goods, culture, and technology, connecting African civilizations with the broader ancient world.

In summary, ancient mining practices in Africa reveal a high level of sophistication and adaptation to the local environment. Early African miners developed effective methods for extracting and processing minerals, which played a crucial role in the economic and cultural development of their societies. These practices not only provided valuable resources but also laid the foundation for Africa's long-standing mining tradition, which continues to be a significant part of the continent's heritage.

<hr>

2. Colonial Era Mining

The colonial era marked a significant turning point in the history of mining in Africa. The arrival of European powers in the late 19th and early 20th centuries led to the systematic exploration and exploitation of the continent's vast mineral resources. This period was characterized by large-scale mining operations, the establishment of new export routes, and profound impacts on the African socio-economic landscape.

European Colonization and Exploitation of Mineral Resources

The Scramble for Africa:

The "Scramble for Africa" in the late 19th century saw European powers vying for control of African territories. Driven by the Industrial Revolution and the need for raw materials, European nations colonized vast regions of Africa, often with little regard for the existing social and political structures. The discovery of valuable minerals played a crucial role in intensifying this scramble.

Exploration and Prospecting:

European explorers and geologists embarked on extensive prospecting missions across the continent. Their goal was to identify and map mineral deposits that could be exploited for the benefit of the colonial powers. This period saw the establishment of numerous mining companies, often backed by European governments and private investors, which set up operations to extract and export these resources.

Exploitation and Labor Systems:

Colonial mining operations were characterized by the exploitation of both natural resources and African labor. The colonial powers imposed harsh labor systems, including forced labor and indentured servitude, to ensure a steady supply of workers for the mines. Africans were often coerced into working under brutal conditions, with little regard for their welfare or rights. The labor-intensive nature of mining, combined with inadequate safety measures, led to high mortality rates and widespread suffering.

Infrastructure Development:

To facilitate the extraction and export of minerals, the colonial powers invested in infrastructure development. Railways, roads, ports, and telegraph lines were constructed to connect mining regions with coastal export points. While these developments were primarily aimed at serving the interests of the colonial economy, they also had long-term impacts on African infrastructure and connectivity.

Economic Impact:

The exploitation of African mineral resources generated immense wealth for the colonial powers, but little of this wealth benefited the local populations. The profits from mining were repatriated to Europe, fueling further industrialization and economic growth there. In contrast, African economies remained largely dependent on the export of raw materials, with limited opportunities for industrialization or economic diversification.

Key Minerals and Their Export Routes

Gold:

Gold was one of the most sought-after minerals during the colonial era. Significant gold deposits were discovered in South Africa's Witwatersrand region in 1886, leading to a gold rush and the rapid development of Johannesburg. South Africa soon became one of the world's largest gold producers. Other notable gold mining regions included Ghana (then the Gold Coast) and Zimbabwe (then Southern Rhodesia).

Diamonds:

The discovery of diamonds in South Africa in the late 19th century transformed the global diamond industry. The Kimberley mine, discovered in 1867, became one of the largest and most famous diamond mines in the world. The De Beers Consolidated Mines Company, founded by Cecil Rhodes in 1888, played a dominant role in the diamond trade, controlling production and prices.

Copper:

Copper mining became a major industry in Central Africa, particularly in the Belgian Congo (now the Democratic Republic of Congo) and Northern Rhodesia (now Zambia). The rich copper deposits of the Katanga region were exploited by companies such as Union Minière du Haut-Katanga, which became one of the world's largest copper producers.

Platinum and Other Metals:

South Africa also emerged as a major producer of platinum, with significant deposits in the Bushveld Complex. Other important minerals included coal, found in abundance in South Africa's Mpumalanga region, and tin, mined extensively in Nigeria and Rwanda.

Export Routes:

To ensure the efficient export of minerals, colonial powers developed extensive transportation networks. Railways were constructed to link mining regions with ports along the coast. For example, the Benguela Railway connected the copper mines of Katanga to the port of Lobito in Angola. Similarly, the Tazara Railway was built to transport copper from Zambia to the port of Dar es Salaam in Tanzania.

Ports played a crucial role in the export of minerals. Major ports included Cape Town and Durban in South Africa, Accra in Ghana, and Mombasa in Kenya. These ports facilitated the shipment of minerals to European markets, where they were processed and utilized in various industries.

Case Studies of Colonial Mining Operations:

1. South Africa:

- The discovery of gold and diamonds in South Africa led to the establishment of large-scale mining operations, transforming the region into a major economic hub. Companies like De Beers and Anglo American played a dominant role in the mining industry, controlling vast mineral resources and shaping the economic landscape of the country.

2. **Congo Free State (Belgian Congo):**

- The exploitation of mineral resources in the Congo Free State, particularly rubber and copper, was marked by extreme brutality and exploitation. King Leopold II of Belgium established a regime of forced labor, resulting in widespread atrocities and significant loss of life. The mining of copper in Katanga became a major economic activity, with the profits enriching the Belgian state and private investors.

3. **Ghana (Gold Coast):**

- Gold mining in Ghana, one of the oldest mining activities on the continent, was intensified during the colonial era. British mining companies established extensive operations, extracting significant quantities of gold for export. The development of railways and ports facilitated the efficient transportation of gold to international markets.

4. **Zimbabwe (Southern Rhodesia):**

- The discovery of gold and other minerals in Zimbabwe led to the establishment of a thriving mining industry. The colonial government and private companies invested in mining infrastructure, contributing to the economic development of the region. However, the benefits of mining were largely confined to the colonial elite and foreign investors.

In summary, the colonial era marked a period of intense exploitation of Africa's mineral resources. European powers established large-scale mining operations, developed extensive export routes, and imposed harsh labor systems on the local populations. While these activities generated immense wealth for the colonial powers, they also had profound and lasting impacts on African societies, economies, and environments. The legacy of colonial mining continues to shape the continent's mining industry and its socio-economic landscape to this day.

3. Post-Colonial Developments

The post-colonial era ushered in significant changes in the mining sector across Africa. With independence movements sweeping the continent in the mid-20th century, newly sovereign nations sought to reclaim control over their natural resources. This period was marked by the nationalization of mines, the implementation of new mining policies, and reforms aimed at fostering economic growth and ensuring that the benefits of mining were more equitably distributed among the population.

Independence Movements and Nationalization of Mines

The Push for Independence:

The mid-20th century was a period of intense political change in Africa, as countries across the continent gained independence from colonial rule. Leaders of the independence movements recognized the strategic importance of their nations' mineral wealth and sought to assert control over these resources as part of their broader efforts to achieve economic self-sufficiency and national sovereignty.

Nationalization of Mines:

One of the key strategies adopted by many newly independent African nations was the nationalization of mining operations. The aim was to transfer ownership and control of mineral resources from foreign companies and colonial governments to the state. This move was seen as essential to ensuring that the profits from mining would be reinvested in the national economy and contribute to development goals.

1. **Zambia:**

- Zambia provides a notable example of this trend. Following independence from Britain in 1964, the Zambian government nationalized the copper mines, which were previously operated by private companies such as Anglo American and the Roan Selection Trust. The nationalization process culminated in the creation of the state-owned Zambia Consolidated Copper Mines (ZCCM) in 1982. The government hoped that nationalization would lead to greater control over the copper sector and increased revenues for national development.

2. **DR Congo**:

- In the Democratic Republic of Congo (formerly Zaire), the nationalization of the mining sector also took place following independence in 1960. The government established the state-owned mining company, Gécamines, to manage the rich mineral resources, particularly copper and cobalt, in the Katanga region. The nationalization aimed to reduce foreign influence and ensure that mining revenues benefited the Congolese people.

3. **Ghana**:

- Ghana, after gaining independence in 1957, also moved towards greater state control over its mineral resources. The government established the State Gold Mining Corporation to oversee gold mining operations. The focus was on increasing state revenue from gold mining and ensuring that the sector contributed to national development.

Challenges and Outcomes:

While nationalization was driven by noble intentions, the outcomes were mixed. Many state-owned mining enterprises struggled with inefficiencies, lack of expertise, and insufficient investment. In some cases, political interference and corruption further hindered the performance of nationalized mines. The initial optimism that nationalization would lead to significant economic benefits often gave way to disillusionment as the anticipated gains failed to materialize.

Major Mining Policies and Reforms

Economic Reforms and Liberalization:

By the late 20th century, many African countries began to shift away from state-led economic models towards more market-oriented approaches. This shift was partly driven by the need to attract foreign investment and expertise to revitalize the mining sector. Economic liberalization and structural adjustment programs, often guided by international financial institutions such as the World Bank and the International Monetary Fund (IMF), led to significant policy reforms in the mining sector.

1. **Privatization and Foreign Investment**:

- Privatization of state-owned mining enterprises became a common strategy to improve efficiency and attract foreign capital. Governments sold stakes in state-owned companies or invited private investors to participate in joint ventures. This shift aimed to bring in new technology, management practices, and investment needed to modernize the mining sector.

2. **Regulatory Frameworks**:

- New regulatory frameworks were introduced to create a more attractive investment climate for foreign mining companies. These frameworks included clearer legal standards, more predictable tax regimes, and better-defined property rights. The goal was to reduce the risks and uncertainties associated with investing in African mining projects.

3. **Environmental and Social Policies**:

- Recognizing the need for sustainable mining practices, many African countries also implemented policies to address environmental and social issues. Environmental impact assessments (EIAs) became a requirement for new mining projects, ensuring that potential environmental damages were identified and mitigated. Additionally, policies aimed at improving labor conditions, community development, and local content requirements were introduced to ensure that mining activities benefited local populations.

Case Studies of Major Policy Reforms:

1. South Africa:

- South Africa, with its extensive mining history, undertook significant reforms in the post-apartheid era. The Mineral and Petroleum Resources Development Act (MPRDA) of 2002 marked a major policy shift, declaring that all mineral resources are the common heritage of all South Africans, with the state as the custodian. The Act aimed to promote equitable access to mineral resources, encourage sustainable development, and address past racial inequalities in the mining sector.

2. Tanzania:

- Tanzania also saw major policy reforms in the mining sector in the late 20th and early 21st centuries. The 1998 Mining Act and subsequent amendments aimed to attract foreign investment by offering incentives such as tax breaks and reduced royalties. However, the government later introduced reforms to increase its share of mining revenues and ensure greater local benefits, including the 2017 Mining Act, which increased royalty rates and required mining companies to list on the local stock exchange.

3. Botswana:

- Botswana is often cited as a success story in terms of mining policies and economic management. Following independence in 1966, the government established Debswana, a joint venture with

De Beers, to manage the country's diamond resources. The partnership ensured substantial revenues for the state, which were reinvested in infrastructure, education, and health. Botswana's stable political environment and prudent economic policies have made it one of the wealthiest countries in Africa.

Impact of Policy Reforms:

The impact of these policy reforms has been significant, though varied across different countries. In many cases, liberalization and regulatory improvements have succeeded in attracting foreign investment and revitalizing the mining sector. However, challenges remain, including ensuring that the benefits of mining are equitably distributed, mitigating environmental impacts, and maintaining stable regulatory environments to sustain investor confidence.

In summary, the post-colonial developments in Africa's mining sector reflect a complex interplay of nationalization, privatization, and policy reform. While the initial wave of nationalization aimed to assert control over natural resources, subsequent reforms have focused on attracting investment, improving regulatory frameworks, and addressing environmental and social concerns. These developments continue to shape the mining industry in Africa, influencing its contribution to economic growth and development.

Chapter 2: Types of Mining in Africa

1. Gold Mining

Gold mining has been a cornerstone of Africa's mining industry for centuries, with the continent being one of the world's leading producers of the precious metal. The allure of gold has attracted prospectors, miners, and investors from around the globe, shaping the economic and social landscape of many African nations.

Major Gold-Producing Countries and Mines

South Africa:

- South Africa has a long history of gold mining and remains one of the largest producers of gold in the world. The Witwatersrand Basin, located in the Gauteng province, is home to some of the deepest and richest gold mines on Earth. Major gold mines in South Africa include the Mponeng, TauTona, and Driefontein mines, which are operated by companies like AngloGold Ashanti and Sibanye-Stillwater.

Ghana:

- Ghana is Africa's second-largest gold producer and has been a significant player in the global gold market for centuries. The country's gold mining industry dates back to the 15th century, with the Ashanti Empire being renowned for its gold wealth. Today, Ghana's gold mines include the Tarkwa and Damang mines operated by Gold Fields, as well as the Ahafo and Akyem mines operated by Newmont.

Mali:

- Mali is another major gold-producing country in Africa, with significant gold deposits located in the western and southern regions of the country. The Sadiola and Morila mines, operated by Barrick Gold and AngloGold Ashanti, respectively, are among Mali's largest gold mines. The country's gold production has been boosted by the development of large-scale mining projects in recent decades.

Tanzania:

- Tanzania's gold mining industry has experienced significant growth in recent years, driven by large-scale mining projects such as the Bulyanhulu and North Mara mines, operated by Barrick Gold. The country's rich mineral reserves and favorable mining policies have attracted substantial investment in the gold sector, making Tanzania a prominent player in Africa's gold market.

Other Countries:

- Other African countries with notable gold mining operations include Burkina Faso, Côte d'Ivoire, Mali, and Zimbabwe, among others. These countries boast significant gold reserves and have attracted investment from multinational mining companies seeking to capitalize on the potential of their gold deposits.

Techniques and Processes

Open-Pit Mining:

- Open-pit mining is a common method used to extract gold from large, near-surface deposits. It involves the excavation of open pits or quarries to access the ore body. Heavy machinery, such as excavators, haul trucks, and drills, is used to remove the overburden and extract the ore. Open-pit mining is suitable for large-scale operations and is often used in conjunction with other mining methods.

Underground Mining:

- Underground mining is employed to access deeper gold deposits that cannot be economically mined using open-pit methods. This technique involves the construction of shafts and tunnels to reach the ore body, where miners extract the gold-bearing rock. Underground mining requires specialized equipment and techniques to ensure the safety of workers and the efficient extraction of ore.

Placer Mining:

- Placer mining, also known as alluvial mining, is a method used to extract gold from riverbeds and stream sediments. It relies on the natural process of erosion and sedimentation to concentrate gold particles in placer deposits. Miners use simple tools such as pans, sluices, and dredges to separate gold from the sand and gravel. Placer mining is often conducted by artisanal miners and small-scale operations.

Heap Leaching:

- Heap leaching is a modern extraction method used to recover gold from low-grade ore or waste rock. It involves stacking ore on a lined pad and applying a leaching solution (usually a cyanide solution) to dissolve the gold. The gold-bearing solution is then collected and processed to recover the gold. Heap leaching is a cost-effective and environmentally friendly method, suitable for large-scale operations with low-grade ore deposits.

Cyanidation:

- Cyanidation, or the cyanide leaching process, is the most widely used method for extracting gold from ore. It involves the dissolution of gold particles in a cyanide solution, followed by the recovery of gold from the solution using activated carbon or zinc precipitation. Cyanidation is highly efficient and can extract gold from very low-grade ore, making it a preferred method for large-scale gold mining operations.

Mercury Amalgamation:

- Mercury amalgamation, although less common today due to its environmental and health risks, was historically used to extract gold from ore. This method involves mixing crushed ore with liquid mercury to form an amalgam, which is then heated to vaporize the mercury and leave behind the gold. Mercury amalgamation is often associated with artisanal and small-scale gold mining, where access to modern extraction methods may be limited.

In summary, gold mining in Africa encompasses a variety of techniques and processes, from large-scale open-pit and underground operations to artisanal and small-scale mining activities. The continent's rich gold deposits have attracted investment from both local and international mining companies, driving economic growth and development in many African countries. However, the industry also faces challenges related to environmental sustainability, social responsibility, and resource management, which must be addressed to ensure the long-term viability of gold mining in Africa.

---

2. Diamond Mining

Diamond mining is a significant industry in Africa, with the continent being home to some of the world's largest diamond deposits. The allure of diamonds has led to the establishment of numerous mines across Africa, contributing to the continent's economic development and global diamond supply.

Key Diamond Mines and Producing Countries

Botswana:

- Botswana is one of the world's leading diamond producers, with diamonds accounting for a significant portion of the country's export revenue. The Jwaneng and Orapa mines, operated by Debswana (a joint venture between De Beers and the Botswana government), are among the largest and richest diamond mines in the world. Botswana's diamond industry has played a crucial role in the country's economic growth and development, providing employment opportunities and government revenue.

Democratic Republic of Congo (DRC):

- The Democratic Republic of Congo is home to significant diamond reserves, particularly in the Kasai region. The country's diamond industry has historically been plagued by conflict and controversy, with diamonds often referred to as "blood diamonds" due to their association with armed conflict and human rights abuses. However, efforts have been made to improve transparency and accountability in the DRC's diamond sector, with initiatives such as the Kimberley Process Certification Scheme aimed at preventing the trade in conflict diamonds.

South Africa:

- South Africa has a long history of diamond mining, dating back to the late 19th century. The country's diamond mines, including the Venetia and Finsch mines, have been instrumental in shaping the global diamond industry. However, South Africa's diamond production has declined in recent years, with other African countries such as Botswana and Angola surpassing it in terms of production.

Angola:

- Angola is another major diamond-producing country in Africa, with significant diamond reserves located in the Lunda Norte and

Lunda Sul provinces. The Catoca mine, operated by a consortium including Endiama (the Angolan national diamond company) and international mining companies, is one of the largest diamond mines in the world. Angola's diamond industry has been a key driver of economic growth and development, although challenges related to transparency and governance remain.

Namibia:

- Namibia is renowned for its marine diamond deposits, which are mined offshore using specialized vessels and equipment. The country's diamond industry is dominated by marine diamond mining operations, with companies like De Beers Marine Namibia and Namdeb Holdings playing a prominent role. Namibia's diamond sector has contributed significantly to the country's economy, providing employment opportunities and government revenue.

Methods of Extraction

Open-Pit Mining:

- Open-pit mining is a common method used to extract diamonds from large, near-surface deposits. It involves the excavation of open pits or quarries to access the diamond-bearing kimberlite or alluvial deposits. Heavy machinery, such as excavators, haul trucks, and bulldozers, is used to remove the overburden and extract the diamond-bearing ore. Open-pit mining is suitable for large-scale diamond deposits and allows for efficient extraction of diamonds.

Underground Mining:

- Underground mining is employed to access deeper diamond deposits that cannot be economically mined using open-pit methods. This technique involves the construction of shafts and tunnels to reach the diamond-bearing kimberlite pipes or veins. Miners extract the diamond-bearing rock using specialized equipment and techniques, ensuring the safety of workers and the efficient extraction of ore. Underground mining is often used in conjunction with open-pit mining to access deeper portions of diamond deposits.

Marine Diamond Mining:

- Marine diamond mining is a specialized method used to extract diamonds from offshore deposits located along the coast or on the seabed. This technique involves the use of specialized vessels equipped with suction hoses, pumps, and processing equipment to extract diamond-rich sediment from the ocean floor. The extracted material is brought to the surface, where it is processed to recover the diamonds. Marine diamond mining is particularly prevalent in countries like Namibia, where significant marine diamond deposits exist.

Alluvial Mining:

- Alluvial mining, also known as placer mining, is a method used to extract diamonds from riverbeds and stream sediments. It relies on the natural process of erosion and sedimentation to concentrate diamonds in placer deposits. Miners use simple tools such as pans, sluices, and dredges to separate the diamonds from the sand and gravel. Alluvial mining is often conducted by artisanal and small-scale miners, particularly in river valleys and alluvial plains.

Artisanal and Small-Scale Mining (ASM):

- Artisanal and small-scale mining (ASM) is a significant component of the diamond mining sector in Africa, particularly in countries with abundant diamond deposits. ASM operations are typically carried out by individuals or small groups using basic equipment and techniques. While ASM provides livelihoods for many people in rural communities, it is often associated with informal practices, environmental degradation, and social challenges.

In summary, diamond mining in Africa encompasses a variety of methods and techniques, from large-scale open-pit and underground operations to marine and alluvial mining activities. The continent's rich diamond deposits have attracted investment from both local and international mining companies, contributing to economic growth and development. However, challenges related to governance, transparency, and sustainability must be addressed to ensure that diamond mining benefits local communities and environments while adhering to responsible mining practices.

3. Other Precious and Base Metals

In addition to gold and diamonds, Africa is rich in other precious and base metals, including platinum, copper, cobalt, and a variety of others. The mining of these metals plays a crucial role in the continent's economy, contributing to export revenue, job creation, and infrastructure development. Let's explore some of the significant metals and their impact on African countries.

Platinum

South Africa:

- South Africa is the world's largest producer of platinum, accounting for a significant portion of global production. The Bushveld Complex in South Africa's North West province is home to extensive platinum deposits, which are mined by companies such as Anglo American Platinum, Impala Platinum, and Sibanye-Stillwater. Platinum mining in South Africa has been a major driver of economic growth and development, providing employment opportunities and export revenue.

Zimbabwe:

- Zimbabwe also has significant platinum reserves, particularly in the Great Dyke region. The country's platinum industry has attracted investment from international mining companies, with projects such as the Unki mine operated by Anglo American Platinum. Zimbabwe's platinum sector has the potential to contribute significantly to the country's economy, although challenges related to governance and investment climate remain.

Other Countries:

- Other African countries with platinum reserves include Russia, South Africa, Zimbabwe, and Botswana. These countries are actively exploring and developing their platinum resources to capitalize on the growing demand for the metal, particularly in industries such as automotive catalytic converters, jewelry, and industrial applications.

Copper

Democratic Republic of Congo (DRC):

- The Democratic Republic of Congo is one of the world's largest producers of copper, with significant copper deposits located in the Katanga region. The country's copper industry has been a key driver of economic growth, providing employment opportunities and export revenue. However, challenges related to governance, infrastructure, and investment climate have hampered the development of the copper sector in the DRC.

Zambia:

- Zambia is another major copper-producing country in Africa, with rich copper deposits located in the Copperbelt region. The country's copper industry has historically been dominated by large-scale mining operations, with companies such as Konkola Copper Mines and Mopani Copper Mines playing a prominent role. Copper mining in Zambia has contributed significantly to the country's economy, although challenges related to taxation, royalty rates, and infrastructure remain.

Other Countries:

- Other African countries with significant copper reserves include South Africa, Botswana, and Namibia. These countries are actively exploring and developing their copper resources to meet the growing demand for the metal, particularly in industries such as construction, electrical wiring, and telecommunications.

Cobalt

Democratic Republic of Congo (DRC):

- The Democratic Republic of Congo is the world's largest producer of cobalt, with the majority of global cobalt reserves located in the country's Katanga region. Cobalt is a critical component in the production of lithium-ion batteries used in electric vehicles and portable electronic devices. However, the cobalt industry in the DRC has faced scrutiny due to concerns about child labor, environmental degradation, and human rights abuses in artisanal and small-scale mining operations.

Other Countries:

- Other African countries with cobalt reserves include Zambia, Morocco, and South Africa. These countries are exploring and developing their cobalt resources to capitalize on the growing demand for battery metals and renewable energy technologies.

Other Metals

Nickel, Iron Ore, and Titanium:

- Africa is also rich in other metals such as nickel, iron ore, and titanium, which are used in various industrial applications. Countries like Madagascar, South Africa, and Sierra Leone have significant reserves of these metals and are actively exploring and developing their resources to meet global demand.

Impact of Mining Operations

Mining operations in Africa have both positive and negative impacts on local communities, economies, and environments. While mining contributes to economic development, job creation, and infrastructure development, it also poses challenges related to environmental degradation, social displacement, and governance.

Positive Impacts:

- Economic Growth: Mining contributes to economic growth and development through export revenue, job creation, and infrastructure development.

- Employment Opportunities: Mining provides employment opportunities for local communities, including both direct employment in mining operations and indirect employment in related industries and services.

- Infrastructure Development: Mining projects often lead to the development of infrastructure such as roads, railways, ports, and power supply, which benefit both mining operations and local communities.

Negative Impacts:

- Environmental Degradation: Mining operations can have significant environmental impacts, including habitat destruction, water pollution, and soil erosion, which can harm local ecosystems and biodiversity.

- Social Displacement: Mining projects can lead to social displacement and conflict, as communities are often displaced from their land to make way for mining operations, leading to loss of livelihoods and cultural heritage.

- Governance Challenges: Mining operations in Africa are often associated with governance challenges such as corruption, lack of transparency, and human rights abuses, which can undermine the benefits of mining for local communities and governments.

In summary, the mining of precious and base metals in Africa plays a crucial role in the continent's economy, contributing to export revenue, job creation, and infrastructure development. While mining offers significant economic opportunities, it also poses challenges related to environmental sustainability, social responsibility, and governance, which must be addressed to ensure that the benefits of mining are equitably distributed and sustainable over the long term.

4. Coal and Fossil Fuels

Coal and fossil fuels have long been essential sources of energy for industrialization and economic development. In Africa, coal mining plays a significant role in meeting the continent's energy needs and driving economic growth. Let's explore the major coal mines and regions in Africa, as well as the role of coal in energy production.

Major Coal Mines and Regions

South Africa:

- South Africa is the largest producer of coal in Africa and one of the world's top coal exporters. The country's coal reserves are primarily located in the Mpumalanga province, with major coal mining operations in the Witbank and Highveld coalfields. Some of the largest coal mines in South Africa include the Goedgevonden, Khutala, and Greenside mines, operated by companies like Anglo American and Exxaro Resources.

Mozambique:

- Mozambique has significant coal reserves located in the Tete province, particularly in the Moatize Basin. The country's coal industry has attracted investment from international mining companies, with projects such as the Moatize mine operated by Vale and the Benga mine operated by Rio Tinto. Mozambique's coal sector has the potential to contribute significantly to the country's economy and energy infrastructure development.

Botswana:

- Botswana is also home to significant coal reserves, particularly in the Morupule Coalfield. The Morupule coal mine, operated by the Morupule Coal Mine (MCM) company, supplies coal to the nearby Morupule Power Station, which is the country's main source of electricity. Botswana's coal industry plays a crucial role in meeting domestic energy demand and supporting economic growth.

Zimbabwe:

- Zimbabwe has coal reserves located in the Hwange region, which are mined by companies such as Hwange Colliery Company Limited. Coal mining in Zimbabwe has historically been a key driver of industrialization and economic development, providing fuel for power generation, manufacturing, and domestic heating.

Other Countries:

- Other African countries with significant coal reserves include Tanzania, Zambia, and Nigeria. These countries are exploring and developing their coal resources to meet domestic energy demand and support economic growth.

Role in Energy Production

Power Generation:

- Coal plays a crucial role in power generation in many African countries, providing a reliable and affordable source of electricity. Coal-fired power plants are commonly used to generate electricity for industrial, commercial, and residential use. In countries like South Africa and Botswana, coal-fired power plants are the backbone of the energy infrastructure, providing a significant portion of the country's electricity supply.

Industrial Use:

- Coal is also used as a fuel in various industrial processes, including steel production, cement manufacturing, and chemical processing. In countries with large coal reserves, such as South Africa and Zimbabwe, coal is an important feedstock for the industrial sector, supporting manufacturing and economic growth.

Domestic Heating:

- In many African countries, coal is used for domestic heating and cooking, particularly in rural areas where access to alternative fuels may be limited. Coal provides a source of heat for households during cold weather and is often used in traditional stoves and fireplaces.

Challenges and Considerations:

- While coal plays an important role in meeting energy needs in Africa, it also poses challenges related to environmental pollution, greenhouse gas emissions, and climate change. The burning of coal releases carbon dioxide and other pollutants into the atmosphere, contributing to air pollution and global warming. As the world transitions towards cleaner and more sustainable energy sources, African countries are increasingly exploring renewable energy alternatives such as solar, wind, and hydropower to diversify their energy mix and reduce reliance on coal.

In summary, coal mining in Africa is a significant industry that provides essential energy resources for industrialization, economic development, and domestic use. While coal remains an important part of the energy mix in many African countries, there is growing recognition of the need to transition towards cleaner and more sustainable energy sources to mitigate environmental impacts and address climate change challenges.

5. Artisanal and Small-Scale Mining

Artisanal and small-scale mining (ASM) is a prevalent form of mining activity in Africa, characterized by its informal nature, small-scale operations, and reliance on rudimentary tools and techniques. Let's explore the definition and scope of ASM, as well as its economic and social significance in African communities.

Definition and Scope

Definition:

- Artisanal and small-scale mining (ASM) refers to mining activities that are conducted on a small scale, often by individuals or small groups, using simple equipment and techniques. ASM is typically characterized by its informal and unregulated nature, with miners often operating outside of formal legal frameworks and without proper licensing or oversight.

Scope:

- ASM encompasses a wide range of mining activities, including the extraction of minerals such as gold, diamonds, gemstones, tin, tantalum, tungsten, and coal. ASM operations can vary in scale from individuals panning for gold in riverbeds to small groups excavating shallow pits or tunnels in search of mineral deposits. ASM is prevalent in rural and remote areas where access to formal employment opportunities may be limited, and mining provides a means of livelihood for local communities.

Economic and Social Significance

Economic Contribution:

- ASM makes a significant contribution to the economies of many African countries, providing employment opportunities, income generation, and poverty alleviation in rural communities. ASM activities often serve as a crucial source of income for individuals and families, particularly in regions where formal employment opportunities are scarce. ASM also contributes to government revenue through taxes, royalties, and fees, although the extent of formal regulation and taxation of ASM varies widely across countries.

Livelihoods and Poverty Alleviation:

- ASM plays a vital role in supporting the livelihoods of millions of people across Africa, particularly in rural and marginalized communities. For many individuals and families, ASM provides a means of subsistence and poverty alleviation, enabling them to meet their basic needs and improve their quality of life. ASM activities often serve as a safety net during times of economic hardship or agricultural downturns, providing alternative sources of income and food security.

Local Development and Empowerment:

- ASM can contribute to local development and empowerment by providing opportunities for entrepreneurship, skills development, and community investment. In many cases, ASM operations are owned and operated by local individuals or communities, allowing them to retain control over their natural resources and benefit directly from mining activities. ASM can also stimulate economic multiplier effects, such as the creation of supporting industries and services, which further contribute to local development.

Challenges and Risks:

- Despite its economic and social significance, ASM also poses numerous challenges and risks, including environmental degradation, health and safety hazards, child labor, and conflict. ASM operations often lack proper infrastructure, equipment, and safety standards, exposing miners to risks such as accidents, injuries, and exposure to hazardous chemicals. Additionally, ASM activities can have negative environmental impacts, such as deforestation, land degradation, and water pollution, which can harm local ecosystems and biodiversity.

In summary, artisanal and small-scale mining (ASM) plays a significant role in Africa's mining sector, providing livelihoods, income generation, and poverty alleviation for millions of people in rural communities. While ASM offers economic opportunities and empowerment for local individuals and communities, it also poses challenges related to environmental sustainability, health and safety, and social responsibility, which must be addressed through improved regulation, support mechanisms, and community engagement.

Chapter 3: Economic Impact of Mining

1. Contribution to National Economies

GDP and Export Revenues:

- The mining sector plays a significant role in contributing to the GDP (Gross Domestic Product) and export revenues of many African countries. Minerals and metals extracted through mining activities are often major export commodities, generating substantial foreign exchange earnings for governments and contributing to overall economic growth. The revenue generated from mining exports can be used to finance government budgets, infrastructure development, and social programs, thus driving broader economic development.

Job Creation and Skills Development:

- Mining activities create employment opportunities across various sectors of the economy, ranging from direct employment in mining operations to indirect employment in supporting industries and services. The mining sector employs a diverse workforce, including engineers, geologists, technicians, machine operators, and administrative staff, providing jobs for both skilled and unskilled workers. Additionally, mining projects often invest in skills development programs and capacity-building initiatives to train local workers and enhance their employability in the mining industry and beyond.

In summary, the mining sector makes significant contributions to national economies in Africa through its impact on GDP, export revenues, job creation, and skills development. By driving economic growth, generating employment opportunities, and fostering skills development, the mining industry plays a crucial role in advancing socioeconomic development and improving livelihoods in African countries.

2. Foreign Direct Investment (FDI)

Investment Trends and Major Investors:

- Foreign Direct Investment (FDI) plays a crucial role in financing mining projects and driving the growth of the mining sector in Africa. Over the years, Africa has attracted significant FDI inflows into its mining industry, with multinational mining companies, private investors, and sovereign wealth funds investing in exploration, development, and production activities. Investment trends in the mining sector are influenced by factors such as commodity prices, regulatory frameworks, political stability, and infrastructure availability.

Role of Multinational Corporations:

- Multinational corporations (MNCs) play a dominant role in the African mining sector, accounting for a substantial portion of FDI inflows and mineral production. These companies bring expertise, technology, and capital to mining projects, contributing to the development of large-scale mining operations and infrastructure. MNCs often operate through joint ventures, partnerships, or subsidiary companies with local governments or private entities, leveraging their global reach and resources to explore and exploit mineral resources in Africa.

In summary, Foreign Direct Investment (FDI) is a key driver of growth and development in the African mining sector, attracting capital, expertise, and technology from multinational corporations and private investors. By financing exploration and development activities, FDI contributes to the discovery and exploitation of mineral resources, generating economic benefits for host countries and supporting sustainable development in the region.

3. Mining and Infrastructure Development

Improvements in Transportation, Energy, and Other Sectors:

- The mining sector often catalyzes infrastructure development in Africa, driving improvements in transportation, energy, and other critical sectors. Mining projects require extensive infrastructure to support exploration, extraction, processing, and exportation activities, leading to investments in roads, railways, ports, power plants, and other essential facilities. Improved transportation infrastructure enhances connectivity between mining sites and markets, reducing logistics costs and facilitating the movement of goods and people. Similarly, investments in energy infrastructure such as power plants and transmission lines provide reliable electricity supply for mining operations and surrounding communities, supporting industrialization and economic growth.

Challenges and Opportunities:

- While mining-related infrastructure development presents significant opportunities for economic growth and development in Africa, it also poses challenges related to financing, governance, and sustainability. The capital-intensive nature of infrastructure projects requires substantial investment from governments, private investors, and international financial institutions, necessitating innovative financing mechanisms and partnerships. Additionally, infrastructure projects must be planned and executed with consideration for environmental sustainability, social impact, and long-term viability, ensuring that they benefit local communities and ecosystems while supporting mining activities. Addressing these challenges requires collaboration between governments, mining companies, civil society organizations, and other stakeholders to develop and implement infrastructure projects that promote inclusive and sustainable development in Africa.

In summary, mining plays a vital role in driving infrastructure development in Africa, leading to improvements in transportation, energy, and other critical sectors. While mining-related infrastructure presents opportunities for economic growth and development, it also requires careful planning, investment, and governance to address challenges and maximize benefits for local communities and the environment. By leveraging infrastructure development to support mining activities and broader socioeconomic development goals, Africa can harness the potential of its mineral resources to promote inclusive and sustainable development across the continent.

Chapter 4: Social Impact of Mining

1. Community Development

Health, Education, and Housing Improvements:

- Mining projects can contribute to community development by supporting improvements in health, education, and housing infrastructure. Companies often invest in healthcare facilities, schools, and housing projects to address the needs of local communities and improve their quality of life. These investments may include the construction of hospitals, clinics, and health centers to provide medical services and support public health initiatives. Similarly, mining companies may build schools, libraries, and vocational training centers to enhance educational opportunities and skills development for community members. Additionally, investments in housing infrastructure such as affordable housing schemes and resettlement programs can help address housing shortages and improve living conditions for residents in mining areas.

Corporate Social Responsibility (CSR) Initiatives:

- Corporate Social Responsibility (CSR) initiatives play a crucial role in promoting community development and addressing social issues associated with mining activities. Mining companies often implement CSR programs to support sustainable development, environmental conservation, and social inclusion in host communities. These initiatives may include community engagement and consultation processes, capacity-building programs, and livelihood enhancement projects aimed at empowering local residents and fostering economic resilience. Additionally, CSR initiatives may involve partnerships with local governments, civil society organizations, and other stakeholders to address community needs and priorities effectively. By integrating CSR principles into their business operations, mining companies can contribute to positive social impacts and build trust and goodwill with local communities.

In summary, community development is an essential aspect of the social impact of mining in Africa, with mining projects often supporting improvements in health, education, and housing infrastructure in host communities. Corporate Social Responsibility (CSR) initiatives play a key role in promoting community development by addressing social issues and fostering inclusive and sustainable development in mining areas. By investing in community development and engaging with local stakeholders, mining companies can contribute to positive social outcomes and build mutually beneficial relationships with host communities.

2. Employment and Labor Conditions

Workforce Demographics and Labor Rights:

- The mining sector in Africa employs a diverse workforce that includes skilled, semi-skilled, and unskilled workers. Employment opportunities in the mining industry can significantly impact local economies by providing jobs and fostering skills development. The demographics of the mining workforce typically reflect the local population, with a mix of men, women, and sometimes migrants from other regions or countries seeking employment.

- Labor rights in the mining sector are critical for ensuring fair treatment, safety, and well-being of workers. Various countries in Africa have established labor laws and regulations to protect workers' rights, including provisions for minimum wage, working hours, health and safety standards, and the right to form or join labor unions. However, the enforcement of these regulations can vary widely, and issues such as inadequate safety measures, poor working conditions, and lack of job security remain challenges in many mining operations.

Issues of Child Labor and Worker Exploitation:

- Child labor is a significant concern in the artisanal and small-scale mining (ASM) sector in Africa. Economic hardship and lack of access to education often drive families to involve their children in mining activities. Child labor in mining is hazardous, exposing children to dangerous working conditions, long hours, and health risks from exposure to toxic substances like mercury and dust. Efforts to combat child labor in mining involve a combination of government regulations, community awareness programs, and initiatives by NGOs and international organizations aimed at providing alternative livelihoods and educational opportunities for children and their families.

- Worker exploitation is another critical issue in the mining sector, particularly in informal and unregulated operations. Exploitation can take various forms, including inadequate wages, excessive working hours, unsafe working conditions, and lack of access to basic services and social protection. Migrant workers and those employed in remote or isolated mining areas are particularly vulnerable to exploitation. Addressing these issues requires a multifaceted approach, including stronger enforcement of labor laws, advocacy for workers' rights, and collaboration between governments, mining companies, and civil society organizations to ensure ethical labor practices and improve working conditions in the mining sector.

In summary, employment and labor conditions in Africa's mining sector present both opportunities and challenges. While the industry provides significant employment opportunities and contributes to local economic development, issues such as child labor, worker exploitation, and inadequate labor rights protection need to be addressed. Ensuring fair labor practices, improving working conditions, and promoting ethical labor standards are essential for fostering a sustainable and socially responsible mining industry in Africa.

3. Cultural and Heritage Considerations
Impact on Indigenous Communities and Heritage Sites:
- Mining activities can have profound impacts on indigenous communities and heritage sites. Indigenous communities often have deep cultural, spiritual, and historical connections to their land, which can be disrupted by mining operations. The extraction of minerals may lead to the displacement of communities, loss of traditional livelihoods, and destruction of culturally significant landscapes and sites. Additionally, mining activities can threaten archaeological sites, sacred grounds, and other heritage locations that hold historical and cultural importance.

- The intrusion of mining operations into indigenous territories can result in social and cultural disintegration, loss of identity, and erosion of traditional knowledge and practices. The introduction of a mining workforce and associated infrastructure can also lead to social changes, such as increased migration, shifts in population dynamics, and potential conflicts between local and migrant communities.

Preservation and Conflict Resolution:

- Preservation of cultural heritage and protection of indigenous rights are crucial components of sustainable mining practices. Mining companies, governments, and other stakeholders must engage in comprehensive cultural heritage assessments and impact studies before initiating mining projects. This involves identifying and mapping culturally significant sites, consulting with indigenous communities, and incorporating their perspectives and knowledge into planning and decision-making processes.

- To mitigate the impact of mining on cultural heritage, companies can adopt measures such as creating buffer zones around heritage sites, relocating important cultural artifacts, and investing in the preservation and restoration of impacted sites. Collaborative efforts with local communities to develop heritage management plans and support cultural programs can also enhance preservation efforts.

- Conflict resolution mechanisms are essential to address disputes and tensions arising from mining activities. This involves establishing clear and transparent processes for community consultation and engagement, ensuring that the rights and interests of indigenous communities are respected and protected. Mechanisms for grievance redressal and dispute resolution should be accessible and responsive, providing communities with avenues to voice their concerns and seek remedies for grievances.

- Governments and mining companies can also implement benefit-sharing arrangements that ensure indigenous communities receive tangible benefits from mining activities, such as revenue-sharing agreements, employment opportunities, and investment in community development projects. These arrangements can help build trust and foster positive relationships between mining companies and indigenous communities.

In summary, the impact of mining on indigenous communities and heritage sites is a critical consideration for sustainable and responsible mining practices in Africa. Preservation of cultural heritage and effective conflict resolution are essential for protecting the rights and interests of indigenous communities and ensuring that mining activities contribute positively to local development. By engaging with indigenous communities, respecting their cultural heritage, and implementing inclusive and transparent decision-making processes, mining companies and governments can promote harmonious and mutually beneficial relationships.

Chapter 5:
Environmental Impact
of Mining

1. Environmental Degradation

Deforestation, Soil Erosion, and Water Pollution:

- Deforestation:

- Mining operations often necessitate the clearing of vast areas of forested land to access mineral deposits, leading to significant deforestation. This deforestation can result in the loss of biodiversity, disruption of ecosystems, and the displacement of wildlife. The removal of vegetation also contributes to increased carbon emissions, exacerbating climate change.

- Soil Erosion:

- The excavation and removal of topsoil during mining activities can lead to severe soil erosion. Without vegetation to anchor the soil, it becomes more susceptible to being washed or blown away, particularly during heavy rains or strong winds. This erosion can reduce the fertility of the land, making it less suitable for agriculture and other land uses post-mining. Additionally, soil erosion can cause sedimentation in nearby rivers and streams, affecting water quality and aquatic habitats.

- Water Pollution:

- Mining activities can result in various forms of water pollution, including acid mine drainage, heavy metal contamination, and increased sediment loads. Acid mine drainage occurs when sulfide minerals exposed during mining react with water and oxygen to form sulfuric acid, which can leach heavy metals into surrounding water bodies. This pollution can have devastating effects on aquatic life, human health, and agricultural productivity. Moreover, the discharge of untreated wastewater and chemicals used in mineral processing can further contaminate water sources, making them unsafe for consumption and use by local communities.

Case Studies of Major Environmental Disasters:

- The Ok Tedi Mine Disaster (Papua New Guinea):

- While not in Africa, the Ok Tedi Mine disaster provides a relevant example of the severe environmental impacts mining can have. In the mid-1980s, tailings and waste rock from the Ok Tedi gold and copper mine were discharged directly into the Ok Tedi River. This led to extensive environmental damage, including the destruction of riverine ecosystems, fish kills, and the displacement of communities reliant on the river for their livelihoods. The long-term environmental degradation has had lasting effects on the biodiversity and health of the affected area.

- The Mufulira Mine (Zambia):

- The Mufulira mine in Zambia has experienced several incidents of sulfur dioxide gas leaks and water contamination. In 2005, a major leak resulted in the discharge of sulfur dioxide gas, affecting the air quality and health of local residents. Additionally, there have been reports of water pollution from the mine's tailings dams, contaminating local water sources with heavy metals and other pollutants. These incidents highlight the ongoing challenges of managing mining waste and preventing environmental contamination.

- The Brumadinho Dam Collapse (Brazil):

- Although not in Africa, the Brumadinho dam collapse in Brazil in 2019 is another example of the catastrophic consequences of mining-related environmental disasters. The collapse of a tailings dam at the Córrego do Feijão iron ore mine released a massive wave of toxic sludge, resulting in the loss of hundreds of lives, extensive environmental damage, and long-term contamination of the surrounding area. This disaster underscores the critical importance of stringent safety standards and effective management of mining waste to prevent similar occurrences in African mining operations.

In summary, mining activities in Africa and globally can lead to severe environmental degradation, including deforestation, soil erosion, and water pollution. Case studies of major environmental disasters highlight the need for stringent environmental regulations, effective management practices, and ongoing monitoring to mitigate the negative impacts of mining and ensure the protection of ecosystems and communities. Addressing these environmental challenges is essential for promoting sustainable mining practices and safeguarding the natural resources and biodiversity of Africa.

2. Sustainable Mining Practices

Rehabilitation and Reclamation Efforts:

- Rehabilitation and Reclamation:

- Rehabilitation and reclamation are critical components of sustainable mining practices aimed at restoring mined land to its natural or economically usable state after mining activities have ceased. This process involves several stages, including the removal of mining infrastructure, reshaping the land, replacing topsoil, and replanting vegetation. Effective rehabilitation plans are designed to minimize environmental impact, promote biodiversity, and support the reintegration of the land into local ecosystems or agricultural production.

- Successful examples of rehabilitation efforts can be seen in countries like South Africa, where the mining industry has implemented comprehensive reclamation projects. For instance, Anglo American's rehabilitated land in the Mpumalanga coal mining region has been transformed into agricultural land, supporting local farming activities and providing a sustainable livelihood for the community. Similarly, in Botswana, the Jwaneng Diamond Mine, operated by Debswana, has ongoing land rehabilitation programs to restore ecological balance and support wildlife conservation.

Green Mining Technologies:

- Energy Efficiency and Renewable Energy:

- Mining operations are typically energy-intensive, relying heavily on fossil fuels for power generation. Green mining technologies aim to reduce the carbon footprint of mining activities by improving energy efficiency and incorporating renewable energy sources. For example, solar and wind power installations are being integrated into mining operations to provide clean and sustainable energy. In South Africa, the South Deep Gold Mine has implemented a solar power plant to reduce its reliance on the national grid and decrease greenhouse gas emissions.

- Water Management and Recycling:

- Efficient water management is crucial for minimizing the environmental impact of mining. Green mining technologies focus on reducing water usage, enhancing water recycling, and preventing water contamination. Techniques such as dry stacking of tailings, where waste materials are dewatered and compacted, can significantly reduce the risk of water pollution. Additionally, the use of advanced water treatment technologies helps ensure that discharged water meets environmental standards. In Namibia, the Husab Uranium Mine employs a comprehensive water recycling system to minimize freshwater consumption and manage wastewater effectively.

- **Waste Reduction and Recycling**:

- Reducing waste generation and promoting the recycling of mining by-products are essential aspects of sustainable mining practices. Technologies such as in-situ leaching, which involves extracting minerals without traditional open-pit or underground mining, can minimize surface disturbance and waste production. Furthermore, reprocessing and recycling tailings and waste rock can recover valuable minerals and reduce the overall environmental footprint of mining operations. The Kansanshi Copper Mine in Zambia has implemented a tailings reprocessing plant that extracts additional copper and gold from previously discarded waste, enhancing resource efficiency and reducing environmental impact.

- **Eco-Friendly Extraction Methods**:

- Innovations in extraction methods aim to reduce the environmental impact of mining by minimizing land disturbance, energy consumption, and chemical use. Bio-mining, for example, utilizes microorganisms to extract metals from ores, offering a less invasive and more sustainable alternative to traditional extraction techniques. Additionally, the development of green chemistry approaches, such as the use of biodegradable chemicals in mineral processing, can reduce the environmental toxicity associated with mining operations.

In summary, sustainable mining practices are essential for mitigating the environmental impact of mining activities and promoting the long-term viability of mining operations in Africa.

Rehabilitation and reclamation efforts, along with the adoption of green mining technologies, play a crucial role in restoring mined land, reducing carbon emissions, managing water resources, minimizing waste, and implementing eco-friendly extraction methods. By integrating these practices, the mining industry can contribute to sustainable development and environmental conservation while continuing to support economic growth and resource extraction.

3. Regulations and Policies

National and International Environmental Regulations:
  - National Environmental Regulations:

- Many African countries have established comprehensive environmental regulations aimed at managing the impact of mining activities on the environment. These regulations typically cover various aspects of mining operations, including land use, water management, air quality, waste disposal, and rehabilitation. For example, South Africa's National Environmental Management Act (NEMA) and the Mineral and Petroleum Resources Development Act (MPRDA) provide a regulatory framework for environmental management in the mining sector, requiring companies to conduct Environmental Impact Assessments (EIAs) and develop Environmental Management Plans (EMPs) before commencing operations.

- In Ghana, the Environmental Protection Agency (EPA) regulates mining activities through the Environmental Assessment Regulations, which mandate EIAs and Environmental Management Plans for all mining projects. Similarly, Tanzania's Environmental Management Act requires mining companies to obtain environmental permits and comply with regulations on waste management, water use, and land rehabilitation.

- **International Environmental Regulations**:

- International regulations and guidelines also play a significant role in shaping environmental practices in the mining sector. Various international bodies and agreements set standards and provide frameworks for environmental protection in mining. The International Finance Corporation (IFC), part of the World Bank Group, has established Performance Standards on Environmental and Social Sustainability, which are widely adopted by mining companies seeking project financing.

- Additionally, the Extractive Industries Transparency Initiative (EITI) promotes transparency and accountability in the management of natural resources, including environmental aspects of mining. The International Council on Mining and Metals (ICMM) sets global standards for sustainable mining practices, emphasizing environmental stewardship, biodiversity conservation, and climate action.

Effectiveness and Enforcement Challenges:

- **Effectiveness of Regulations**:

- The effectiveness of environmental regulations in the mining sector depends on their design, implementation, and enforcement. Well-designed regulations that are clear, comprehensive, and aligned with international best practices can significantly mitigate the environmental impact of mining activities. Effective regulations also require robust monitoring and reporting mechanisms to ensure compliance and track environmental performance over time.

- However, the effectiveness of these regulations varies across countries and regions, influenced by factors such as institutional capacity, political will, and stakeholder engagement. In some cases, mining companies have successfully integrated regulatory requirements into their operations, leading to improved environmental outcomes and sustainable practices.

- **Enforcement Challenges**:

- Enforcement of environmental regulations in the mining sector faces several challenges, including limited resources, inadequate capacity, and corruption. Regulatory agencies in many African countries may lack the financial and human resources needed to conduct regular inspections, monitor compliance, and enforce penalties for violations. This can result in weak oversight and limited accountability for environmental breaches.

- Additionally, political and economic pressures can undermine enforcement efforts. Governments may prioritize economic growth and revenue generation from mining over strict environmental regulation, leading to regulatory leniency or selective enforcement. Corruption and lack of transparency can further complicate enforcement, allowing companies to evade compliance through bribes or political influence.

- To address these challenges, strengthening institutional capacity, enhancing transparency, and fostering collaboration between governments, industry, and civil society are essential. Building the technical expertise of regulatory agencies, increasing funding for environmental monitoring and enforcement, and implementing anti-corruption measures can improve regulatory effectiveness. Engaging local communities and civil society organizations in monitoring and advocacy can also enhance accountability and ensure that environmental regulations are enforced effectively.

In summary, national and international environmental regulations play a critical role in managing the impact of mining activities and promoting sustainable practices in the African mining sector. While these regulations have the potential to significantly mitigate environmental degradation, their effectiveness depends on robust design, implementation, and enforcement. Addressing enforcement challenges through capacity building, transparency, and stakeholder engagement is essential for ensuring that mining operations adhere to environmental standards and contribute to sustainable development.

Chapter 6: Case Studies of Major Mining Countries

1. South Africa

Historical and Contemporary Mining Landscape:

- Historical Context:

- South Africa's mining history dates back over a century and is deeply intertwined with its economic and social development. The discovery of diamonds in Kimberley in 1867 and gold in the Witwatersrand in 1886 marked the beginning of large-scale mining in the country. These discoveries led to a mining boom, attracting international investors and migrant labor, and establishing South Africa as a major player in the global mining industry. The mining sector played a pivotal role in the country's industrialization, infrastructure development, and urbanization.

- During the apartheid era, the mining industry was characterized by racial segregation and labor exploitation. Black South African miners faced harsh working conditions, low wages, and limited rights. The struggle for labor rights and better working conditions in the mining sector became a significant aspect of the broader anti-apartheid movement.

- Contemporary Landscape:

- Today, South Africa remains one of the world's leading mining countries, with a diversified mining industry that includes gold, platinum, coal, diamonds, and a variety of other minerals. The country's mining sector continues to be a major contributor to its economy, accounting for a significant portion of GDP, export revenues, and employment. However, the industry also faces several contemporary challenges, including labor disputes, regulatory changes, and the need for transformation and sustainability.

- Post-apartheid, there have been significant efforts to address the historical injustices and promote inclusive growth within the mining sector. Policies such as the Mining Charter aim to ensure that the benefits of mining are more equitably distributed, promoting black economic empowerment, skills development, and community investment.

Key Minerals and Their Global Significance:

- **Gold:**

- South Africa has historically been one of the largest gold producers in the world, with the Witwatersrand Basin being the epicenter of its gold mining industry. Although production has declined in recent years due to the depletion of easily accessible reserves and increased production costs, South Africa remains an important player in the global gold market. Gold mining has significant economic implications, contributing to foreign exchange earnings and employment.

- **Platinum Group Metals (PGMs):**

- South Africa holds the largest reserves of platinum group metals (PGMs), including platinum, palladium, rhodium, and others. The Bushveld Complex is the world's most important source of these metals, which are critical for various industrial applications, including automotive catalytic converters, electronics, and jewelry.

South Africa's dominance in the PGM market makes it a key supplier globally, influencing market prices and industrial demand.

- **Diamonds**:

- The discovery of diamonds in Kimberley was a catalyst for the mining industry in South Africa. The country remains a significant diamond producer, with major operations in the Kimberley, Venetia, and Cullinan mines. South African diamonds are known for their quality and contribute substantially to the country's export revenues and global market presence.

- **Coal**:

- Coal is another major mineral resource in South Africa, primarily used for electricity generation and as an export commodity. The country has some of the largest coal reserves in the world, and the industry plays a crucial role in its energy sector. However, coal mining also poses environmental challenges, particularly in terms of greenhouse gas emissions and water usage.

- **Other Minerals**:

- South Africa is also a significant producer of manganese, chromium, and vanadium, all of which are essential for the global steel industry. The country's rich mineral diversity supports a wide range of industrial activities and contributes to its strategic importance in the global mining sector.

In summary, South Africa's mining sector has a rich historical legacy and continues to play a pivotal role in its economy. The country is a leading producer of key minerals with significant global implications, including gold, platinum group metals, diamonds, and coal. While the industry faces contemporary challenges, it remains a critical component of South Africa's economic landscape, driving growth, employment, and international trade.

2. **Democratic Republic of Congo (DRC)**

Rich Mineral Resources and Associated Conflicts:

- Mineral Wealth:

- The Democratic Republic of Congo (DRC) is one of the most mineral-rich countries in the world, possessing vast reserves of copper, cobalt, gold, diamonds, tin, tantalum, and tungsten. The country's mineral wealth, particularly in the Katanga region, has the potential to drive significant economic growth and development. The DRC holds some of the largest deposits of cobalt and copper globally, which are critical for the production of batteries, electronics, and renewable energy technologies. This makes the DRC a strategic player in the global supply chain for these essential minerals.

- Conflict Minerals:

- Despite its mineral wealth, the DRC has been plagued by conflicts associated with the mining sector. The term "conflict minerals" refers to minerals that are mined in conditions of armed conflict and human rights abuses, particularly in the eastern regions of the DRC. Rebel groups and militias have exploited mineral resources to fund their activities, leading to prolonged violence, instability, and humanitarian crises. The trade in conflict minerals has been linked to severe human rights violations, including forced labor, child labor, and sexual violence.

- International efforts to address the issue of conflict minerals have led to the implementation of regulations and initiatives such as the Dodd-Frank Act in the United States, which requires companies to disclose the use of conflict minerals sourced from the DRC and surrounding countries. The European Union has also introduced regulations aimed at ensuring responsible sourcing of minerals.

These efforts seek to reduce the flow of funds to armed groups and promote transparent and ethical supply chains.

Economic and Social Challenges:

- Economic Challenges:

- The DRC's mining sector has the potential to generate significant revenue and drive economic growth. However, the country faces numerous economic challenges that hinder the realization of this potential. Corruption, mismanagement, and lack of infrastructure are major impediments to effective resource management and development. The benefits of mining often fail to reach local communities due to inadequate governance and transparency.

- The informal and artisanal mining sectors, which employ a significant portion of the population, are characterized by unsafe working conditions, lack of regulation, and low wages. Artisanal miners often work in hazardous conditions with little access to modern tools or safety equipment, exposing them to health and safety risks.

- Social Challenges:

- Social challenges in the DRC's mining sector include widespread poverty, human rights abuses, and lack of social services. The exploitation of mineral resources has not translated into improved living conditions for the majority of the population. Instead, communities in mining areas often experience environmental degradation, displacement, and social unrest.

- Child labor is a critical issue in the DRC's mining sector, particularly in artisanal and small-scale mining. Children are often employed in dangerous tasks, such as digging and carrying heavy loads, which can have severe physical and psychological impacts. Efforts to combat child labor include international partnerships, community awareness programs, and initiatives aimed at providing education and alternative livelihoods for children and their families.

- Environmental degradation is another significant challenge. Mining activities, especially artisanal mining, can lead to deforestation, soil erosion, and water pollution. The use of toxic chemicals such as mercury in gold mining poses serious health risks to miners and local communities and can have long-term environmental consequences.

In summary, the Democratic Republic of Congo is endowed with vast mineral resources that have the potential to drive economic development and prosperity. However, the country faces significant challenges, including conflicts associated with mineral exploitation, economic mismanagement, and severe social issues such as poverty, human rights abuses, and environmental degradation. Addressing these challenges requires comprehensive reforms, effective governance, and international cooperation to ensure that the DRC's mineral wealth benefits its population and contributes to sustainable development.

3. **Botswana**

Successful Diamond Mining Industry:

- **Historical Development:**

- Botswana's diamond mining industry began in earnest in the 1960s, with the discovery of significant diamond deposits in Orapa in 1967. This marked a turning point for the country, which was one of the poorest in the world at the time of its independence in 1966. The establishment of the Orapa mine, followed by the discovery and development of the Jwaneng and Letlhakane mines, positioned Botswana as a leading diamond producer.

- **Major Mines and Production**:
- Botswana is home to some of the richest diamond mines globally. The Jwaneng mine, often referred to as the richest diamond mine in the world by value, produces a substantial portion of Botswana's diamond output. Other significant mines include Orapa, the second-largest diamond-producing mine in the world by area, and the Letlhakane and Damtshaa mines.
- The production from these mines has made Botswana one of the largest diamond producers by value. The diamonds mined in Botswana are known for their high quality, which has secured the country a strong position in the global diamond market.
- **Partnership with De Beers**:
- A critical factor in the success of Botswana's diamond industry is the partnership between the government of Botswana and De Beers, a leading diamond company. This partnership, known as Debswana, is a 50-50 joint venture and has been instrumental in ensuring the effective management and profitability of Botswana's diamond resources. Debswana operates the major diamond mines and has significantly contributed to the country's economic stability and growth.

Economic Transformation and Governance:
- **Economic Growth**:
- The revenue generated from diamond mining has been pivotal in transforming Botswana's economy. Diamonds account for a significant portion of the country's GDP, export earnings, and government revenue. The prudent management of diamond revenues has facilitated substantial investments in infrastructure, education, health, and social services, contributing to one of the highest standards of living in Africa.

- The government has implemented sound fiscal policies, saving a portion of diamond revenues in sovereign wealth funds, such as the Pula Fund, to ensure long-term economic stability and intergenerational equity. This approach has helped Botswana avoid the "resource curse" that has plagued many other resource-rich countries.

- **Good Governance and Transparency**:

- Botswana is often cited as a model of good governance and transparency in the management of natural resources. The government has established a stable political environment, characterized by democratic governance, respect for the rule of law, and low levels of corruption. Transparency in the allocation of mining rights, revenue management, and public spending has fostered trust and confidence among investors and citizens alike.

- Institutions such as the Botswana Unified Revenue Service (BURS) ensure effective collection and management of mining revenues, while the Ministry of Mineral Resources, Green Technology, and Energy Security oversees the regulation and development of the mining sector. Regular audits and public disclosures contribute to accountability and transparency in the sector.

- **Diversification Efforts**:

- Recognizing the finite nature of diamond resources, Botswana has embarked on economic diversification efforts to reduce dependence on diamonds. The government has invested in other sectors such as tourism, agriculture, manufacturing, and financial services to broaden the economic base. Initiatives like the Economic Diversification Drive (EDD) aim to support local industries and create employment opportunities beyond the mining sector.

- Additionally, Botswana is focusing on beneficiation and value addition within the diamond industry itself. The establishment of cutting and polishing facilities, as well as diamond trading centers like the Diamond Trading Company Botswana (DTCB), seeks to capture more value within the country before diamonds are exported.

In summary, Botswana's successful diamond mining industry has been a cornerstone of its economic transformation, driving growth, development, and improved living standards. The strategic partnership with De Beers, combined with sound governance, prudent fiscal management, and transparency, has ensured that diamond revenues are effectively utilized for national development. Efforts to diversify the economy and add value to the diamond sector further enhance Botswana's resilience and long-term economic prospects.

4. Ghana

Gold Mining and Its Economic Contributions:

- Historical Context:

- Ghana, often referred to as the "Gold Coast" during colonial times, has a long history of gold mining dating back to ancient times. The country's gold mining history is a central part of its cultural and economic identity, with significant gold production starting in the pre-colonial era. The arrival of Europeans in the 15th century led to increased gold trade, which continued to expand over the centuries.

- Major Gold Mines and Production:

- Ghana is one of the largest gold producers in Africa and the world. Key gold mining regions include Ashanti, Western, and Central regions. Major gold mines include the Obuasi Mine, operated by AngloGold Ashanti, the Tarkwa Mine, operated by Gold Fields, and the Ahafo and Akyem mines, operated by

Newmont. These mines contribute significantly to the country's gold production, which accounts for a substantial portion of its export revenues and GDP.

- The Ashanti Goldfields Corporation, now part of AngloGold Ashanti, has historically been one of the major gold mining companies in Ghana. The Obuasi Mine, in particular, is one of the richest gold mines in the world and has been a significant contributor to Ghana's gold production.

- **Economic Contributions**:

- Gold mining is a major economic driver in Ghana, contributing significantly to the national economy. The sector accounts for a substantial share of the country's GDP, export earnings, and employment. Gold exports are a primary source of foreign exchange, which is crucial for economic stability and development.

- The mining industry also stimulates other sectors of the economy, such as construction, manufacturing, and services, through demand for goods and services. Additionally, the government earns significant revenue from mining through taxes, royalties, and dividends, which are essential for funding public services and infrastructure development.

Environmental and Social Issues:

- **Environmental Degradation**:

- Despite its economic benefits, gold mining in Ghana has significant environmental impacts. The use of mercury and cyanide in gold processing poses serious risks to water bodies, soil, and human health. Mercury contamination, in particular, affects aquatic life and can enter the food chain, posing long-term health risks to local communities.

- Deforestation and land degradation are also major environmental concerns. Large-scale mining activities often result in the clearing of forests and agricultural land, leading to habitat loss and reduced biodiversity. Open-pit mining and the improper disposal of tailings can cause soil erosion and sedimentation in rivers, affecting water quality and availability.

- In response to these challenges, the government and mining companies have implemented various environmental management practices, including land reclamation, reforestation, and improved waste management. However, enforcement of environmental regulations remains a challenge, particularly in artisanal and small-scale mining (ASM) operations.

- **Social Issues**:

- The social impact of gold mining in Ghana includes both positive and negative aspects. On the positive side, mining provides employment opportunities and supports local economies. Large-scale mining companies often invest in community development projects, including education, healthcare, and infrastructure, as part of their corporate social responsibility (CSR) initiatives.

- However, the sector also faces significant social challenges. Artisanal and small-scale mining, which employs a large number of people, often operates informally and without adequate safety measures. This leads to hazardous working conditions, accidents, and health issues for miners. Child labor is another critical issue in the ASM sector, with children engaged in dangerous tasks that can have detrimental effects on their health and education.

- Land displacement and conflicts over land use are also prevalent, particularly in regions where mining activities encroach on agricultural land and communities. Disputes between mining companies and local communities can arise over issues such as compensation, environmental degradation, and access to resources.

- Efforts to address these social issues include government regulations to formalize and regulate ASM, initiatives to eliminate child labor, and community engagement programs by mining companies. NGOs and international organizations also play a role in advocating for the rights and welfare of affected communities.

In summary, gold mining is a critical component of Ghana's economy, providing substantial economic benefits through revenue generation, employment, and foreign exchange earnings. However, the sector also poses significant environmental and social challenges, including pollution, deforestation, hazardous working conditions, and social conflicts. Addressing these issues requires a combination of effective regulation, corporate responsibility, and community engagement to ensure that the benefits of gold mining are maximized while minimizing its negative impacts.

Chapter 7: Challenges and Controversies

1. Resource Curse and Economic Diversification

Dependence on Mining Revenues:

- Resource Curse Phenomenon:

- Many resource-rich countries, particularly in Africa, have experienced the paradoxical phenomenon known as the "resource curse." This term refers to the negative consequences associated with abundant natural resource wealth, including economic volatility, governance challenges, social unrest, and stagnated development.

- One of the key manifestations of the resource curse is the overreliance on revenues from extractive industries, such as mining, which can distort the economy, undermine other sectors, and foster corruption and rent-seeking behavior. Dependence on mining revenues exposes countries to fluctuations in commodity prices, geopolitical risks, and environmental degradation without necessarily translating into sustainable development or broad-based prosperity.

- Fiscal Pressures and Macroeconomic Vulnerabilities:

- Countries heavily reliant on mining revenues may experience fiscal pressures and macroeconomic vulnerabilities due to the volatility of commodity markets. Windfall profits during periods of high commodity prices may lead to overheating of the economy, inflationary pressures, and currency appreciation, making other sectors less competitive. Conversely, downturns in commodity prices can result in revenue shortfalls, budget deficits, and economic downturns, exacerbating social inequalities and poverty.

Strategies for Economic Diversification:

- **Reducing Dependence on Mining**:

- Economic diversification is essential for mitigating the risks associated with the resource curse and building resilient economies. Governments of resource-rich countries need to pursue policies that promote diversification away from reliance on extractive industries towards more sustainable and inclusive economic sectors.

- **Investing in Infrastructure and Human Capital**:

- Investing in infrastructure, education, and healthcare is crucial for laying the foundations for economic diversification. Improved infrastructure, such as transport networks, energy systems, and telecommunications, facilitates the development of other industries and enhances competitiveness. Investing in education and skills development creates a more skilled workforce capable of contributing to diverse sectors of the economy.

- **Promoting Industrialization and Value Addition**:

- Promoting industrialization and value addition is another key strategy for economic diversification. Instead of exporting raw materials, countries should focus on processing and adding value to their natural resources domestically. This not only creates more value and employment opportunities but also reduces vulnerability to commodity price fluctuations.

- **Supporting Small and Medium-Sized Enterprises (SMEs)**:

- Supporting the growth of small and medium-sized enterprises (SMEs) across various sectors can drive economic diversification and innovation. SMEs are often more adaptable to changing market conditions and can contribute to job creation, poverty reduction, and inclusive growth. Governments can provide targeted support through access to finance, technical assistance, and regulatory reforms to facilitate SME development.

- **Promoting Sustainable Development Goals (SDGs)**:

- Aligning economic diversification efforts with the Sustainable Development Goals (SDGs) provides a framework for inclusive and sustainable development. Pursuing diversification strategies that prioritize environmental sustainability, social inclusion, and resilience can help countries achieve long-term development objectives while reducing their dependence on extractive industries.

In summary, addressing the challenges associated with the resource curse and promoting economic diversification are essential for fostering sustainable and inclusive development in resource-rich countries. By reducing dependence on mining revenues and investing in infrastructure, human capital, industrialization, and SME development, countries can build resilient economies that are less vulnerable to external shocks and better equipped to achieve long-term prosperity.

2. **Corruption and Governance Issues**

Corruption in the Mining Sector:

- **Prevalence and Impact**:

- Corruption is a pervasive issue in the mining sector of many countries, undermining transparency, accountability, and equitable distribution of benefits. In resource-rich countries, the lucrative nature of mining activities often creates opportunities for corruption, rent-seeking behavior, and illicit financial flows.

- Corruption in the mining sector can take various forms, including bribery, embezzlement, nepotism, and collusion between government officials, mining companies, and other stakeholders. It distorts decision-making processes, favors vested interests over public welfare, and erodes trust in institutions.

- **Key Challenges**:

- One of the key challenges is the lack of transparency and accountability in the awarding of mining licenses, permits, and contracts. Non-transparent licensing processes can facilitate rent-seeking behavior and elite capture, allowing powerful individuals or companies to exploit mineral resources at the expense of the public interest.

- Weak regulatory oversight and enforcement mechanisms exacerbate corruption risks in the mining sector. Inadequate monitoring, limited capacity, and regulatory capture can create opportunities for regulatory arbitrage and illicit activities, such as illegal mining, tax evasion, and environmental violations.

Transparency Initiatives and Anti-Corruption Measures:

- **Promoting Transparency and Accountability**:

- Transparency initiatives play a crucial role in combating corruption and improving governance in the mining sector. Enhanced transparency enables stakeholders, including citizens, civil society organizations, and investors, to access information about mining contracts, revenues, payments, and environmental and social impacts.

- Governments can promote transparency through legal and regulatory reforms that require companies to disclose relevant information, such as beneficial ownership, revenue payments, and environmental compliance. Implementing mechanisms for public participation, such as stakeholder consultations and access to information laws, enhances accountability and oversight.

- **EITI and Extractive Sector Transparency**:

- The Extractive Industries Transparency Initiative (EITI) is a global standard for promoting transparency and accountability in the extractive sector, including mining. EITI requires member countries to disclose information about the management of natural resources, including revenues, contracts, and payments, in a publicly accessible and comprehensible manner.

- By adhering to EITI standards, countries can enhance transparency, build trust between government, industry, and civil society, and reduce corruption risks in the mining sector. EITI implementation involves multi-stakeholder engagement and regular reporting, facilitating dialogue and collaboration among diverse actors.

- Anti-Corruption Measures and Legal Frameworks:

- Strengthening anti-corruption measures and legal frameworks is essential for combating corruption in the mining sector. Governments can enact and enforce laws that criminalize corruption, bribery, and other illicit activities, with appropriate penalties for offenders.

- Establishing independent anti-corruption agencies and specialized anti-corruption units within regulatory bodies can enhance enforcement capacity and promote accountability. Whistleblower protection mechanisms and reporting channels enable individuals to expose corrupt practices without fear of retaliation.

- Corporate Integrity and Compliance:

- Mining companies also play a crucial role in combating corruption by adhering to ethical business practices, promoting integrity, and implementing robust anti-corruption compliance programs. Adopting codes of conduct, conducting due diligence on business partners, and implementing internal controls and monitoring mechanisms can help companies prevent and detect corrupt practices within their operations.

In summary, addressing corruption and governance issues in the mining sector requires a multi-faceted approach involving government reforms, transparency initiatives, anti-corruption measures, and corporate integrity efforts. By promoting transparency, accountability, and ethical behavior, countries can strengthen governance frameworks, reduce corruption risks, and foster sustainable and responsible mining practices that benefit society as a whole.

3. Conflict Minerals

Definition and Examples:

- **Conflict Minerals**:

- Conflict minerals are natural resources, typically minerals or metals, whose extraction and trade are linked to armed conflict, human rights abuses, and violence. The mining and trade of conflict minerals often fund armed groups and contribute to conflict, instability, and humanitarian crises in affected regions.

- **Examples**:

- The most commonly known conflict minerals include:

- **Tantalum**: Used in electronic devices such as smartphones and laptops, tantalum is sourced from minerals like coltan, primarily mined in the Democratic Republic of Congo (DRC) and surrounding countries in Central Africa.

- **Tin**: Used in electronics, tin is sourced from minerals like cassiterite, which is also mined in conflict-affected areas in Central Africa.

- **Tungsten**: Used in various industries, including aerospace and defense, tungsten is sourced from minerals like wolframite, which is often mined in conflict zones.

- **Gold**: A widely used precious metal in jewelry, electronics, and finance, gold mining in conflict-affected areas can contribute to armed conflict, human rights abuses, and environmental degradation.

International Efforts to Address the Issue:

- Regulatory Measures:

- Several regulatory measures have been enacted to address the issue of conflict minerals and promote responsible sourcing practices. The most notable of these is Section 1502 of the Dodd-Frank Wall Street Reform and Consumer Protection Act in the United States. This legislation requires companies listed on U.S. stock exchanges to disclose whether their products contain conflict minerals originating from the DRC or adjoining countries. The aim is to increase transparency and accountability in supply chains and reduce the funding of armed groups through mineral trade.

- Certification Programs:

- Certification programs such as the Conflict-Free Smelter Program (CFSP) and the Responsible Minerals Assurance Process (RMAP) aim to certify smelters and refiners of conflict minerals based on responsible sourcing practices. These programs provide assurances to downstream companies and consumers that the minerals used in their products have been sourced ethically and responsibly.

- International Initiatives:

- International initiatives such as the Kimberley Process Certification Scheme (KPCS) and the Extractive Industries Transparency Initiative (EITI) also play a role in addressing the issue of conflict minerals. The Kimberley Process focuses on preventing the trade in conflict diamonds, while EITI promotes transparency and accountability in the extractive sector, including mining.

- **Corporate Due Diligence**:

- Many companies have implemented due diligence processes to trace the origin of minerals in their supply chains and ensure that they are not sourcing from conflict-affected areas. This involves engaging with suppliers, conducting risk assessments, and implementing controls to prevent the use of conflict minerals in their products.

- **Civil Society and Advocacy**:

- Civil society organizations, non-governmental organizations (NGOs), and advocacy groups play a crucial role in raising awareness about the issue of conflict minerals, advocating for policy reforms, and pressuring companies to adopt responsible sourcing practices. Campaigns such as "Conflict-Free" aim to mobilize consumers to demand ethically sourced products and hold companies accountable for their supply chains.

In summary, conflict minerals pose significant challenges due to their links to armed conflict, human rights abuses, and environmental degradation. International efforts to address the issue include regulatory measures, certification programs, international initiatives, corporate due diligence, and civil society advocacy. By promoting responsible sourcing practices and increasing transparency in mineral supply chains, stakeholders aim to mitigate the negative impacts of conflict minerals and contribute to peace, stability, and sustainable development in affected regions.

Chapter 8: The Future of Mining in Africa

1. Technological Innovations

Advances in Mining Technologies:

- Automation and Robotics:

- Rapid advancements in automation, robotics, and artificial intelligence (AI) are transforming the mining industry by enabling remote operation, autonomous vehicles, and predictive maintenance. Automated drilling, hauling, and processing systems enhance operational efficiency, safety, and productivity while reducing reliance on manual labor.

- Data Analytics and AI:

- Data analytics and AI technologies are revolutionizing decision-making processes in mining operations. By analyzing vast amounts of data from sensors, drones, and geological surveys, mining companies can optimize exploration, extraction, and processing processes, leading to higher yields, lower costs, and reduced environmental impact.

- Advanced Exploration Techniques:

- Advanced exploration techniques, such as satellite imaging, remote sensing, and geophysical surveys, are providing unprecedented insights into mineral deposits and geological structures. High-resolution imagery and geological modeling tools enable more accurate targeting of exploration activities, reducing exploration risks and costs.

Potential for Increased Efficiency and Sustainability:

- **Operational Efficiency**:

- Technological innovations offer significant opportunities for improving operational efficiency and resource utilization in the mining sector. By optimizing workflows, reducing downtime, and streamlining processes, mining companies can increase productivity, minimize waste, and enhance profitability.

- **Environmental Sustainability**:

- Sustainable mining practices are becoming increasingly important as concerns about environmental degradation, climate change, and resource depletion grow. Technological innovations, such as green mining technologies, renewable energy solutions, and water recycling systems, can help minimize the environmental footprint of mining operations and mitigate negative impacts on ecosystems and local communities.

- **Social Responsibility**:

- The future of mining in Africa will also be shaped by efforts to promote social responsibility and community engagement. Mining companies are increasingly recognizing the importance of building strong relationships with local communities, respecting human rights, and contributing to sustainable development through corporate social responsibility (CSR) initiatives, community development projects, and inclusive stakeholder engagement processes.

Conclusion:

- The future of mining in Africa is poised for significant transformation driven by technological innovations, increased efficiency, and sustainability considerations. By embracing advanced technologies, mining companies can unlock new opportunities for growth, improve operational performance, and minimize environmental and social impacts. However, realizing the full potential of these innovations will require collaboration, investment, and a commitment to responsible and sustainable mining practices that benefit all stakeholders.

2. **Renewable Energy and Mining**

Integration of Renewable Energy in Mining Operations:

- **Solar Power:**

- Solar power is increasingly being integrated into mining operations across Africa, offering a reliable and sustainable energy source for remote and off-grid sites. Solar photovoltaic (PV) systems are deployed to power mining equipment, lighting, and infrastructure, reducing dependence on diesel generators and grid electricity.

- **Wind Power:**

- Wind power presents another viable option for mining companies to diversify their energy sources and reduce reliance on fossil fuels. Wind turbines can be installed on-site or in nearby wind-rich areas to generate clean electricity for mining operations, providing a cost-effective and environmentally friendly alternative to traditional energy sources.

- **Hydropower:**

- Hydropower, harnessing the energy of flowing water, is well-suited for mining operations located near rivers, streams, or waterfalls. Small-scale hydropower systems can provide a continuous and renewable source of electricity for mining activities, with minimal environmental impact compared to large-scale hydroelectric dams.

  - **Biomass and Biofuels**:

  - Biomass and biofuels derived from organic waste or agricultural residues offer potential for powering mining operations in regions with abundant biomass resources. Bioenergy technologies, such as biogas digesters and biomass boilers, can convert organic materials into heat and electricity, providing a sustainable energy solution for mining activities.

  Potential for Reducing Carbon Footprint:

  - **Emissions Reduction**:

  - The integration of renewable energy sources in mining operations presents a significant opportunity for reducing carbon emissions and mitigating climate change. By replacing fossil fuel-based energy sources with renewables, mining companies can lower their greenhouse gas emissions and contribute to global efforts to limit global warming and transition to a low-carbon economy.

  - **Energy Efficiency**:

  - In addition to renewable energy adoption, mining companies are implementing energy efficiency measures to optimize energy consumption and reduce their carbon footprint. Technologies such as energy-efficient equipment, process optimization, and waste heat recovery systems help minimize energy wastage and improve overall operational efficiency.

  - **Carbon Neutrality Goals**:

  - Many mining companies are setting ambitious carbon neutrality goals and adopting strategies to achieve net-zero carbon emissions across their operations. This includes investing in renewable energy projects, offsetting emissions through carbon credits or reforestation initiatives, and implementing sustainable practices throughout the mining value chain.

Conclusion:

- The integration of renewable energy in mining operations offers a dual opportunity for reducing carbon emissions and enhancing sustainability. By leveraging solar, wind, hydropower, and bioenergy resources, mining companies can transition towards cleaner and more resilient energy systems while minimizing environmental impacts and contributing to climate change mitigation efforts. Embracing renewable energy technologies represents a crucial step towards a more sustainable future for the mining industry in Africa and beyond.

3. Policy and Regulatory Developments

Emerging Policies and Their Potential Impact:

- Local Content Requirements:

- Many African countries are implementing local content requirements to promote domestic participation and benefit sharing in the mining sector. These policies mandate mining companies to source goods and services locally, employ local labor, and invest in skills development and technology transfer. While local content requirements aim to maximize the socio-economic benefits of mining activities, they also pose challenges such as capacity constraints, technology gaps, and regulatory compliance.

- Environmental Regulations:

- Strengthening environmental regulations is a key priority for many African governments to address the environmental impacts of mining operations. New policies may include stricter environmental standards, increased monitoring and enforcement, and requirements for environmental impact assessments and remediation. By enhancing environmental regulations, governments seek to minimize pollution, protect ecosystems, and ensure sustainable resource management.

- **Community Engagement and Consent**:

- Recognizing the importance of community engagement and consent, emerging policies emphasize the rights of local communities affected by mining activities. Governments are implementing legal frameworks that require mining companies to consult with communities, obtain free, prior, and informed consent (FPIC), and share benefits equitably. These policies aim to prevent social conflicts, promote inclusive development, and safeguard the rights of indigenous peoples and marginalized groups.

Role of International Organizations and Agreements:

- **Multilateral Organizations**:

- Multilateral organizations such as the World Bank, International Monetary Fund (IMF), and United Nations (UN) play a significant role in shaping mining policies and regulations in Africa. Through technical assistance, capacity building, and policy advice, these organizations support governments in developing sustainable mining frameworks, promoting good governance, and addressing social and environmental challenges.

- **Regional Agreements**:

- Regional agreements and initiatives also influence mining policies and regulatory frameworks in Africa. Organizations such as the African Union (AU) and regional economic communities (RECs) collaborate to harmonize mining laws, facilitate cross-border cooperation, and promote regional integration. Regional agreements may address issues such as resource management, infrastructure development, and investment promotion to enhance the competitiveness and sustainability of the mining sector.

- **International Standards and Guidelines**:

- International standards and guidelines, such as the Extractive Industries Transparency Initiative (EITI), the Kimberley Process Certification Scheme (KPCS), and the United Nations Guiding Principles on Business and Human Rights, provide frameworks for responsible and transparent mining practices. By adhering to these standards, governments, companies, and civil society organizations can enhance accountability, promote ethical conduct, and respect human rights in the mining sector.

Conclusion:

- Emerging policies and regulatory developments in the mining sector in Africa reflect a growing recognition of the need for sustainable and responsible mining practices. By addressing issues such as local content, environmental protection, and community engagement, governments seek to maximize the socio-economic benefits of mining while minimizing its negative impacts. International organizations and agreements play a crucial role in supporting governments, fostering cooperation, and promoting best practices to ensure that the mining sector contributes to inclusive and sustainable development in Africa.

Conclusion

1. Summary of Key Points

Throughout this book, we have explored various aspects of mining in Africa, examining its historical context, current practices, and future prospects. Here's a recap of the major themes and findings:

- **Historical Context**: We delved into the rich history of mining in Africa, from ancient civilizations to colonial exploitation and post-colonial developments. Understanding this history provides valuable insights into the present-day dynamics of the mining sector.

- **Types of Mining**: We explored the diverse array of minerals mined in Africa, including gold, diamonds, base metals, coal, and fossil fuels. Each type of mining has its unique characteristics, challenges, and contributions to the economy and society.

- **Economic Impact**: Mining plays a significant role in Africa's economy, contributing to GDP, export revenues, and job creation. However, it also poses challenges such as dependence on volatile commodity prices and the need for sustainable development.

- **Social Impact**: The social impact of mining is multifaceted, with both positive and negative aspects. While mining creates employment opportunities and supports community development, it also raises concerns about labor conditions, human rights, and cultural heritage preservation.

- **Environmental Impact**: Mining activities can have profound environmental consequences, including deforestation, water pollution, and habitat destruction. Addressing these impacts requires robust regulations, sustainable practices, and technological innovations.

- **Challenges and Controversies**: We discussed challenges such as the resource curse, corruption, conflict minerals, and governance issues that affect the mining sector in Africa. These challenges highlight the importance of effective governance, transparency, and accountability in the industry.

- **The Future of Mining**: Looking ahead, we explored the potential of technological innovations, renewable energy integration, and policy developments to shape the future of mining in Africa. Embracing sustainable practices and responsible governance will be crucial for realizing the sector's full potential.

In conclusion, mining in Africa is a complex and dynamic industry with profound implications for the continent's economy, society, and environment. By addressing its challenges and leveraging its opportunities, African countries can harness the benefits of mining while mitigating its negative impacts, ultimately contributing to inclusive and sustainable development for generations to come.

2. Final Thoughts

As we contemplate the future of mining in Africa, it is essential to reflect on the opportunities and challenges that lie ahead. While the sector has historically been a significant driver of economic growth and development, its sustainability and inclusivity are increasingly coming into focus.

Reflections on the Future: The future of mining in Africa holds promise as technological innovations, renewable energy integration, and evolving policies reshape the industry landscape. With the right strategies and investments, mining can continue to play a vital role in driving economic prosperity, creating employment opportunities, and fostering innovation.

Potential for Sustainable and Inclusive Growth: Sustainable and inclusive growth in the mining sector is not only desirable but also achievable. By embracing responsible mining practices, promoting environmental stewardship, and engaging with local communities, mining companies can contribute positively to the socio-economic development of Africa while minimizing negative impacts.

Key Considerations: As we envision the future of mining in Africa, several key considerations emerge. These include:

- **Environmental Sustainability**: Prioritizing environmental sustainability is paramount to safeguarding natural resources, ecosystems, and biodiversity for future generations.

- **Social Responsibility**: Upholding social responsibility entails respecting human rights, promoting community development, and ensuring equitable benefit-sharing from mining activities.

- **Governance and Transparency**: Strengthening governance frameworks, enhancing transparency, and combating corruption are essential for building trust, fostering investor confidence, and maximizing the benefits of mining for society.

Collaborative Efforts: Achieving sustainable and inclusive growth in the mining sector requires collaborative efforts from governments, mining companies, civil society organizations, and international stakeholders. By working together towards common goals, we can overcome challenges, seize opportunities, and unlock the full potential of mining as a catalyst for development.

In conclusion, the future of mining in Africa holds immense potential for sustainable and inclusive growth. By embracing innovation, promoting responsible practices, and fostering collaboration, we can harness the transformative power of mining to build a brighter and more prosperous future for Africa and its people.

References

1. World Bank. (2020). "The African Mineral Sector: Opportunities and Challenges." Retrieved from https://www.worldbank.org/en/region/afr/publication/the-african-mineral-sector-opportunities-and-challenges

2. United Nations Economic Commission for Africa. (2019). "Africa Mining Vision." Retrieved from https://www.uneca.org/

sites/default/files/PublicationFiles/
amv_eng_revised_miningvision.pdf

3. International Institute for Sustainable Development. (2021). "Mining in Africa: Policy Perspectives." Retrieved from https://www.iisd.org/sites/default/files/publications/mining-africa-policy-perspectives.pdf

4. Extractive Industries Transparency Initiative. (2020). "EITI in Africa: Key Figures and Trends." Retrieved from https://eiti.org/sites/default/files/documents/
eiti_in_africa_key_figures_and_trends_0.pdf

5. United Nations Development Programme. (2018). "African Mining Vision: Progress and Achievements." Retrieved from https://www.africanminingvision.org/resource/african-mining-vision-progress-and-achievements/

6. International Council on Mining and Metals. (2020). "Mining: Partnering for Sustainable Development." Retrieved from https://www.icmm.com/website/publications/pdfs/mining-partnering-for-sustainable-development

7. World Economic Forum. (2021). "The Future of Mining and Metals in Africa." Retrieved from https://www.weforum.org/agenda/2021/02/future-of-mining-metals-africa/

8. African Development Bank Group. (2019). "Mineral Resource Governance in the 21st Century." Retrieved from https://www.afdb.org/sites/default/files/documents/publications/mineral-resource-governance-21st-century-44800.pdf

9. International Monetary Fund. (2020). "Africa's Mining Sector: A Blessing or a Curse?" Retrieved from https://www.imf.org/en/News/Articles/2020/09/29/na092920-africas-mining-sector-a-blessing-or-a-curse

10. United Nations Environment Programme. (2021). "Global Trends in Artisanal and Small-Scale Mining (ASM): A Review of Key Numbers and Issues." Retrieved from https://wedocs.unep.org/bitstream/handle/20.500.11822/35198/ASM_Global%20Trends%20in%20Artisanal%20and%20Small-Scale%20M

Title: Mining in Africa: Past, Present, and Future

Description:

"Mining in Africa: Past, Present, and Future" offers a comprehensive exploration of the rich tapestry of the mining industry on the African continent. From ancient civilizations to modern-day developments, this book delves into the historical context, economic significance, social impacts, and environmental challenges of mining in Africa.

Readers will embark on a journey through time, tracing the evolution of mining practices from traditional methods to advanced technologies. They will discover the diverse array of minerals mined in Africa, including gold, diamonds, base metals, coal, and fossil fuels, and explore the economic, social, and environmental implications of each.

Drawing on a wealth of research and analysis, this book examines the role of mining in Africa's economy, its contribution to GDP, export revenues, job creation, and infrastructure development. It also highlights the challenges and controversies facing the mining sector, from the resource curse and corruption to conflict minerals and environmental degradation.

Looking ahead, "Mining in Africa: Past, Present, and Future" offers insights into the future of mining on the continent. It explores emerging trends, technological innovations, renewable energy integration, and policy developments shaping the trajectory of the industry. With a focus on sustainability, inclusivity, and responsible governance, this book presents a roadmap for realizing the full potential of mining as a catalyst for development in Africa.